The Ether Dispute

**Revisiting Einstein's argument
that space is a physical substance**

By Richard J. Wilson J.D.

Published 2010
ISBN 9781452817910

Contact author at: theether@cox.net

Table of Contents

"By far the greatest hindrance and aberration of the human understanding proceeds from the dullness, incompetency, and deception of the senses; in that things which strike the sense outweigh things which do not immediately strike it, though they are more important. Hence it is that speculation commonly ceases where sight ceases; insomuch that of things invisible there is little or no observation."

Francis Bacon: *Novum Organum 1620*

Introduction

Einstein

"We may say that according to the general theory of relativity space is endowed with physical qualities; in this sense, therefore, there exists ether. Space without ether is unthinkable; for in such empty space there would be no propagation of light."

Einstein: *Leyden Address* 1920

The Dispute

There is a serious dispute by very intelligent people over the physical nature of space. The dispute arises because of what Francis Bacon called the "dullness, incompetence, and deceptions" of our innate human senses." When we observe outer space, our human senses lead us to two different conclusions:

On the one hand, if we judge by appearance, space appears and feels to us like an empty void, leading some to conclude the earth and heavenly bodies are spinning in an empty vacuum. On the other hand, if we judge by *behavior,* we find space exhibits physical behaviors – it carries light and heat from the sun and stars – leading others to conclude space is some sort of invisible substance called "the ether."

Parties to the Dispute: Cosmology champions the conclusion that space is an empty vacuum. Although often confused with astrophysics, cosmology is not a modern science like astrophysics. It does not use the scientific method. It is a scholastic philosophy that uses the ancient scholastic method of inquiry, and seeks the ancient goal described in Aristotle's Greek *Organum.*

Cosmologists divide the universe into abstract categories according to appearance to arrive at the conclusion that the universe is composed of the base elements *of space, energy,* and *matter.* It then uses logic to speculate upon *where* the universe came from, *where* it will end up, and *how many* universes exist. It then gathers all evidence supporting the logical conclusion, while ignoring any evidence that conflicts.

Early biblical cosmologists using this method and seeking this goal developed the theory that in the beginning there was only God and a totally empty void. Lonely God then miraculously created earth and heavenly bodies to float in the void, and created man to populate the earth and keep Him company. They also speculated there are two other invisible universes called Heaven and Hell where people go after life in this universe, and predicted, if man doesn't mend his ways, God will destroy our universe in a fiery apocalypse.

Modern cosmologists, because they use the same ancient method of inquiry and seek the same goal, naturally arrive at the same conclusion as biblical cosmologists. They theorize our universe began as an empty vacuum and that a mysterious explosion created the earth and heavenly bodies that are now speeding into the empty void. They also speculate there are probably many alternative universes, and predict that ours may someday collapse in a fiery apocalypse.

Because they leave out the concept of God, modern cosmologists would have us believe their Big Bang Theory, as it is commonly called, is the result of advances in modern science that just happen to agree with Genesis. But Francis Bacon observed in his *Novum Organum* that they were already promoting a theory based on Genesis in 1620 before advanced technology and use of the scientific method:

"Some moderns have with extreme levity indulged so far as to attempt to found a system of natural philosophy based on the first chapter of Genesis . . . and bring them into the view of the world so fashioned and masked, as if they were complete in all parts and finished."

On the other hand, astrophysics champions the conclusion that space is some sort of invisible physical substance. Using the modern scientific method of inquiry formally introduced in 1620 by Bacon's *Novum Organum,* astrophysicists divide the universe into physical parts according to their observed behavior, and seek to determine *what* the universe is made of, *how* it produces its behaviors, and *the laws* that nature follows to govern the behavior.

Using this different method and seeking this different goal, astrophysicists theorize that – since space, energy, and matter all exhibit physical behaviors – they must all three be a physical part of a unified field of interrelated substances that will eventually be found to be interconnected and interchangeable. They call this the Unified Field Theory and it was how Einstein viewed the universe.

As we can see, the Unified Field Theory views space differently from the Big Bang Theory of cosmology. Cosmologists refer to space as "an empty vacuum," and astrophysicists refer to it as "the ether." However, remembering the Galileo incident, both parties were careful to ignore their differences and quietly pursue their own inquiry with their own methods.

And our establishment, also remembering the incident, carefully published papers in its forums from both qualified cosmologists and qualified astrophysicists. Thus the dispute over the nature of space simmered quietly for 400 years with both sides patiently awaiting further evidence from advancing technology to resolve their differences.

Dispute Brought Progress: We need to note that the theory of modern cosmology has been substantially complete since biblical days, but, while it satisfied our longing to know *where* the universe came from, the theory proved sterile of progress. It didn't lead us to discover *what* the universe is made of or *how* it creates its behaviors. We knew as little about the universe before Copernicus than did ancient biblical cosmologists.

Francis Bacon made these points in his *Novum Organum.* He noted that scholastic philosophies like cosmology seeking *where* things came from leads to discourse, but no works or progress. And he argued this lack of progress should be taken for a sign of its sterility and the need for a new method of inquiry:

> *"They (scholastics) make the quiescent principles* **wherefrom***, and not the moving principles* **whereby** *things are produced, the object of the contemplation and inquiry. The former tend to discourse, the latter to works...*
>
> *"Fruits and works are sponsors and sureties for the truth of philosophies. For what is founded on nature grows and increases; while what is founded on opinion varies but increases not...*
>
> *"It is idle to expect any great advancement in science from the super inducing and engrafting of new things upon old. We must begin anew from the very foundations, unless we would revolve forever in a circle with mean and contemptible progress."*

And Bacon proved correct. Progress in understanding our universe only began when astrophysicists like Copernicus, Galileo, and Kepler decided to ignore Genesis, and study the *behavior* of our universe to determine *what* it is made of and *how* it creates its behaviors. This information led Newton to discover the physical laws that govern its behavior to set the stage for our modern Space Age; and led Einstein to develop Relativity to set the stage for our modern Atomic Age.

So, while the Big Bang theory of cosmology and Genesis and the conclusion that space is an empty vacuum proved sterile for thousands of years; the new conclusion of astrophysics that space is an invisible substance and part of a Unified Field interrelated with energy and matter led to progress and our modern world.

Dispute Victim of Cold War: Unfortunately, toleration by the establishment for the view of astrophysics ended during the "red scare" of the Cold War. Rumors began to circulate that the Unified Field Theory of astrophysics that contradicted Genesis was a product of "godless communism" designed to disrupt the Christian west; and, as the record shows, our establishment lost its cool and reverted to the same intolerant behavior that went on in the infamous Galileo incindent.

Wanting nothing to do with anything even remotely rumored to be connected with "godless communism," the establishment banned from its forums any further discussion of the Unified Field Theory, and any reference to space as "the ether." Since then papers accepted by establishment forums must agree with the Big Bang Theory patterned on Genesis, and refer to space as "an empty vacuum," or be summarily rejected. And, today, 20 years after the Cold War ended, the ban remains.

Thus, all the work of genius like that of Newton, Faraday, Poincare, Lorentz and Einstein that led to so much progress and our modern society has been swept off the table. Our establishment is now serving as a Defender of the Faith, protecting the Big Bang story of Genesis, and behaving in the same medieval manner Bacon observed in 1620 when all establishment forums were dominated by scholastic philosophy:

"In the customs and instructions of schools, academies, colleges, and similar bodies destined for the abode of learned men and the cultivation of learning, everything is found adverse to the progress of science. For the lectures and exercises are so ordered that to think or speculate on anything out of the common way can hardly occur to any man.

"And if one or two have the boldness to use any liberty of judgment, they must undertake the task all by themselves; they can have no advantage from the company of others. And if they can endure this also, they will find their industry and largeness of mind no slight hindrance to their fortune.

"For the studies of men in these places are confined and as it were imprisoned in the writing of certain authors, from whom, if any man dissent, he is straightway arraigned as a turbulent person and an innovator."

Of course, cosmologists would have us believe there is a scientific basis for the ban. They argue the Michelson and Morley experiment of 1884 proved "scientifically" that space is empty, and that no such thing as the ether exists. And, for good measure, they add that Einstein's *Theory of Relativity* supports the M and M finding. This argument is used by establishment forums as the excuse for its medieval behavior.

However, the argument is bogus. The fact is the modern Unified Field Theory continued to be developed by astrophysicists like Poincare, Lorentz and Einstein *after* the M and M experiment, and Einstein made a special effort in his 1920 *Leyden Lecture* to mention the experiment, and to state categorically that Relativity "does not deny the ether." Yet the establishment goes on perpetuating the bogus argument.

Grounds for the Indictment: It may seem ridiculous to accuse our establishment of such medieval behavior. After all America is the leader of our modern Space Age. It landed man on the moon, and is still busy exploring our solar system with space travel and the space telescope. But careful notice will reveal we're drifting along today on technology and information about the behavior of the universe that is at least a half-century old.

13

This drifting since the beginning of The Cold War can be seen in two ways:

First, relying on the Big Bang Theory, we're not making breakthrough discoveries in our space programs as we should. Cosmology has already determined *where* the universe came from, *where* it is going to end up, *how many* alternative universes probably exist, and concluded that space is an empty vacuum. With the theory complete, our space voyages merely seek to polish Genesis. So, gaining little from our space program, we've begun cutting the budget and repeating the same behavior Bacon tells us existed in 1620:

> *"The logic now in use serves rather to fix and give stability to the errors which have their foundation in commonly received notions than to help search after truth. So it does more harm than good. . .*
>
> *"For if you look at the method of them and divisions, they seem to embrace everything which can belong to the subject, and although these divisions are ill filled out, to the common mind they present the form and plan of a perfect science . . .*
>
> *"It is nothing strange if men do not seek to advance in things delivered to them as long since perfect and complete."*

Secondly, we can see the drifting in the books and papers published under the auspices of our establishment in the last half-century. They, too, are nothing more than a mere polishing of the sterile Big Bang story of Genesis. Each book or paper may include some recently discovered evidence – *provided it supports the biblical story of the creation and beginning of time* – but carefully ignores any new evidence that conflicts with the biblical story.

A survey of the books and papers will reveal we're no longer seeking *what* the universe is made of or *how* the universe produces relativity, inertia, gravity or transmits light and heat – questions that raged in establishment forums before astrophysics was banned a half century ago. Establishment supported books and papers now follow the same tired pattern Bacon noted in 1620:

"For let a man look carefully into all the variety of books, he will find endless repetitions of the same thing, varying in the method of treatment, but not new in substance, insomuch, what was a question once is a question still, and instead of being resolved by discussion, it is only fixed and fed."

Of course, there are books published outside the establishment arguing the theory of astrophysics. There is Eric Lerner's *The Big Bang Never Happened* arguing space is populated by plasma: There is Walter Isaacson's best-selling *Einstein: His life and universe* that devotes a full chapter to the work Einstein did on the Unified Field theory, and of his support of "the ether" concept: And there is Professor Jane Gregory's *Fred Hoyle's Universe* reporting how Hoyle was professionally destroyed for daring to challenge the Big Bang story.

But, like all books that conflict with Genesis, such books are either totally ignored by our establishment, or, if noted at all, dammed with faint praise. Meanwhile, all the new information about the physical behaviors of space reported by advancing technology over the last 50 years lies ignored and unexamined. Our establishment hasn't bothered to collect and catalogue the evidence, much less study the patterns it produces.

We Are Playing with Fire: If Francis Bacon were alive today, he'd warn us we're committing exactly the same error Italy and Spain did in his 17th century. Free discussion had led Spain and Italy to give birth to *The Age of Exploration*, and the new *Renaissance* of science; but, at the very height of their success, as a result of the Galileo incident, the two nations suddenly banned any further discussion of the universe in establishment forums that conflicted with biblical cosmology and punished anyone who persisted.

The tragic result was leadership of the age, and all the wealth and power it could have provided Spain and Italy, passed to northern countries where the ban was not in effect. Those nations took over the exploration of the world, and the renaissance of science, and enjoyed the political power and wealth it brought, while poor Italy and Spain, sticking with the sterile theory of the universe of biblical cosmology, wasted for the lack of progress for centuries.

Bacon would remind us there are emerging nations today that have no interest in Genesis; but that are surely reading and discussing Newton and Einstein. Without any ban upon astrophysics in their establishment forums, they may very well be studying the accumulating evidence that we are ignoring, and, like Italy and Spain, we may awake one day to discover they have snatched our leadership and begun enjoying all the wealth and power it brings, leaving us in the dust.

The Purpose of This Work: This work is an attempt to avoid a repeat of that tragic scenario. We'll begin in Part I by collecting and cataloging the new evidence of the physical behaviors of space provided by advancing technology, and see how it tends to support the Unified Field Theory and the view that space is a physical substance related to

17

energy and matter. I think many readers will find the catalog interesting enough to study in detail and, perhaps, make some constructive additions.

But, since this is not a scientific treatise designed to settle the ether dispute, merely an argument for reinstating it, be sure to read enough of Part I to convince yourself that Newton, Faraday, Poincare, Lorentz and Einstein were not fools to conclude space is a physical substance and part of a Unified Field, and, when satisfied, skip down to the second part.

In Part II we'll explore exactly how our establishment came to the medieval decision to ban discussion of the Unified Field Theory and the ether dispute that these famous astrophysicists spent their lives developing. Then in Part III we'll study a proven way to get astrophysics and its theories reinstated in establishment forums to preserve our place in The Space Age and the wealth and power it provides us all.

Part I

Isaac Newton

"That one body may act upon another at a distance through a vacuum without the mediation of anything else, by and through which their action and force may be conveyed from one to another, is to me so great an absurdity that, I believe, no man who has a competent faculty for thinking could ever fall into it."

Isaac Newton: *Notebooks*

A Partial Catalogue of The Physical Behaviors of Space

From Newton to Einstein, astrophysicists worked practically blind. None saw galaxies or pictures of beautiful clouds in outer space in the definition provided by our modern space telescope or space travel. Telescopes in their time barely penetrated our atmosphere, and the exploration of space was still in the future. Yet, based upon the fact that space supported the transmission of light and heat from the sun, and supported the peculiar behaviors of gravity and inertia, they concluded that space could not be an empty vacuum, but must be some sort of invisible physical *atmosphere* they called the ether.

Genius like Newton was not alone in observing that space acts like it is a physical atmosphere. One English country gentleman wrote Newton that, unless space has a physical atmosphere to resist the effect of gravity, everything would fall into the closest large star, and the whole universe soon end up in one giant ball. Newton agreed, circulated the letter, and entered similar thoughts in his notebook to formally begin the ether dispute.

Evidence from advancing technology like the space telescope and observations during space travel tends to confirm this view.

Space Behaves Like
A Physical Atmosphere

Let's begin by noting that both space and earth's atmosphere are totally invisible to the human senses. We can't tell where one begins and the other ends. The only way to determine that

20

earth's atmosphere is a physical substance is to notice how it affects the behavior of objects. We see objects in earth's atmosphere behave in certain physical patterns. Now advancing technology provides clear evidence that objects in outer space also behave in the same physical patterns.

For example, if we compare photos of spiral galaxies in space with satellite photos of hurricanes in earth's atmosphere – photos available to anyone on the internet – without a caption we'd be hard put to tell which is formed in space and which in earth's atmosphere. Both spiral galaxies and hurricanes have the same eye in the center, the same giant windswept spiral arms, and both move slowly along sweeping up everything in their path.

And compare space telescope photos of the beautiful billowing clouds hovering in the distant Nebula of Orion, with the beautiful billowing clouds that hover in earth's atmosphere over Arizona and New Mexico during monsoon season. Again, it's hard to tell which formed in space and which formed in earth's atmosphere. Both are large billowing clouds of physical substance reflecting the light of their neighboring star.

And compare space photos of tornados in Kansas sucking up everything in their path, the kind that transported Dorothy to the Land of Oz; with astronomer drawings of black holes in space that appear to be sucking up everything around them, transporting the stuff to somewhere as mysterious as Oz. They look like twins.

In general, if we examine the navigation charts NASA publishes for each trip into outer space, we find them strikingly similar to navigation charts made for voyages in earth's oceans and atmosphere. All of them chart physical impediments that need to be avoided, plan the same maneuvers to accelerate speed, and specify the same conditions of timing. Let's single out a few of the most common similarities:

Tides and Currents: Like earth's oceans and atmosphere, space has tides and currents. Every voyage in space is timed to begin according to the position of the moon and sun to take advantage of the physical tides in space created by them, just as ships captains wait for the right position of moon and sun for tides to assist them to launch voyages in earth's oceans.

And, after launch, the spacecraft will move into some known current in space created by a physical mass to assist the craft to sail more easily across open space, just as a ship or an airplane will drop into an ocean current or jet stream to assist them. So we find physical tides and currents are as much a part of the topography of outer space as in earth's oceans and atmosphere.

Winds: There are winds in outer space similar to those in earth's atmosphere. When a comet comes close to the sun, the comet's million-mile smoke-like tail doesn't point toward the sun as the law of gravity would dictate. The tail points away from the sun. Cosmology, assuming space is an empty vacuum, has concluded that something called "plasma" is emitted by the sun into the empty vacuum of space to blow the tails away.

However, it is just as likely that the sun superheats the surrounding substance of space, and the hot substance of space turns into plasma to carry the particles of the comet's tail away from the hot sun by convection, just like the superheated atmosphere around a fire carries particles of smoke up and away from a fire in defiance of gravity. And the effects of solar winds can be detected deep into space.

Waves: There are also physical waves in space similar to the tsunami waves in earth's oceans. Whenever a supernova explodes in outer space, the violence creates very destructive shockwaves that flow across space for billions of miles. And radio operators report that when the sun has explosive flares, it sends out tsunami like shockwaves into space that interfere with radio signals. It is also assumed that an atom bomb exploded in space will send out physical shockwaves powerful enough to change the course of dangerous asteroids.

Whirlpools: As noted before, astronomers using the NASA space telescope report that space contains whirlpools called black holes that suck-up everything in their path – just like whirlpools in oceans and tornadoes in the atmosphere suck up everything in their path. But there is also evidence that space has smaller "keyholes" or rough spots or turbulence that can interfere with the course of a spacecraft, just as earth's oceans and atmosphere have rough spots or turbulence that can interfere with the course of a ship or airplane.

Noise: Space has static noise similar to that in earth's atmosphere. We've long known disturbances in earth's atmosphere create radio static, and now supersensitive modern infrared receivers have detected a similar ubiquitous static coming from all directions and every aspect of space. Apparently all of the physical behaviors of space produce radio static just as all the physical behaviors in earth's atmosphere create radio static.

Friction: According to Einstein, objects traveling through space are subject to friction similar to objects traveling through earth's atmosphere and oceans in proportion to the size of the object and its speed. He theorized that the friction will increase until the object reaches the speed of light, and then it would be so strong the object could accelerate no further because there isn't enough energy in the universe to overcome the friction.

And we all know space, like earth's atmosphere and oceans, provides resistance that reduces the amplitude of vibrations of heat and light over distance until they reach a point where their amplitudes are so weak the heat and light can no longer be detected. The only difference between friction in space and in our atmosphere and oceans is the intensity. The friction is much stronger in earth's atmosphere and oceans, and therefore much more noticeable.

Distortions: Space physically distorts just like earth's atmosphere and oceans. Whenever earth's atmosphere or oceans are distorted by heat or explosion, or an area is displaced by invasion of a large object, they develop pressures so strong they can push over huge buildings and sink huge ships.

Likewise, whenever space is distorted by heat or explosion, or an area is displaced by large objects, space develops pressure so strong it pushes large objects about or holds them in orbit.

However, there is a difference. When earth's atmosphere or oceans are distorted, we can feel the pressure on our skin, but when space is distorted we can't feel the pressure on our skin. All we feel is physically heavier or lighter as we pass through a gravitational distortion. This leads some to believe, since we feel nothing on our skin and can see nothing, gravitational pressure is due to magical rays emanating from a nearby object.

But it's possible we feel nothing on our skin because space doesn't act at the molecular level, but acts *only* at the atomic level on every atom of our body equally. Since the pressure is evenly distributed over every atom of our body, we feel nothing except heavier and lighter as we move through a gravitational field. This is an important observation because it could tell us something about the physical nature of space.

And distortion of space can also be created by magnets. If we sprinkle iron filings on a piece of paper and move a magnet nearby, the iron filings will neatly arrange into the distortions created by the magnet. And, while we think our bodies don't react to magnetic distortions, medical science disagrees. Doctors now prescribe magnets to speed healing of our bones after an operation.

Furthermore, we can observe that cows often line up north and south in a field, and birds and whales appear to use earth's magnetic distortions to migrate. And we humans can use a magnetized needle floating in water to line up with the distortions of space created by earth's magnetism to migrate as well.

And there are more distortions of space than produced by gravity and magnetism. As noted before, a sudden explosion, whether in space or earth's atmosphere or oceans, creates a distortion of the atmosphere around the explosion that dramatically affects objects that get in the way.

Vibrations: We've known for a long time that what we hear as sound is merely a vibration of air or water molecules in earth's atmosphere and oceans, and that we can send out these vibrations and use the echo bouncing off objects to navigate like bats in the dark. This fact convinced 17th century scientists that earth's atmosphere must be an invisible physical stuff, and ultimately they discovered what it was made of and how it creates the behavior.

Likewise, we've known for some time that space can carry vibrations of light and heat that can be bounced off objects to navigate about in space with radar. This fact convinces astrophysicists that space must also be a physical substance that could eventually be understood, and the knowledge used like we use knowledge of earth's atmosphere.

Barriers: Space also provides a physical barrier to the speed of light and heat vibrations just like earth's atmosphere and oceans provide a physical barrier to the speed of sound and heat vibrations. Earth's atmosphere limits sound vibrations to 760 feet per second; while outer space limits the speed of light vibrations to 186,200 miles per second; and both produce the same Doppler Effect. So, regarding barriers to the speed of vibrations; space and earth's atmosphere and oceans behave the same.

Energy: Space and earth's atmosphere are not mere passive mediums transporting light and heat from the sun; they both provide innate energy. Earth's atmosphere holds up clouds, drives atmospheric pumps, and resists movement of objects. Likewise the substance of space forcefully resists any object that changes direction or speed. Unlike the force of gravity, there's nothing we can blame this force of inertia upon except energy provided by the substance of space itself.

Human Sensations: Despite common belief, astronauts sense very much everything in outer space as they do in earth's atmosphere. Of course, they sense no molecules of air or water brushing against their bodies in space and can't breathe in space because space has no particles of air or water. However they sense everything else. In fact a clever pilot can maneuver a jet plane in earth's atmosphere to cause a blindfolded astronaut to think he or she is in outer space.

Astronauts see the same light and feel the same heat of the sun in space as in earth's atmosphere, and they feel the same effects of inertia and gravity. And when their spacecraft fires jets of energy to push off against the surrounding space, they feel the same acceleration a jet pilot feels when he fires his aircraft engine to push off against earth's atmosphere, or a squid feels when it uses its jets of water to push off in earth's ocean.

Of course, they can't use the same wings or rudder they use in earth's atmosphere to gain lift and guide the spacecraft in outer space; but this may be because the wings and rudder are simply not big enough or the craft lacks sufficient speed to deform space as the atmosphere. However, if the wings and rudder were the size of the moon, or the craft were traveling near the speed of light, they might well deform space to provide the lift and guidance.

Helter-Skelter: Space also exhibits the same helter-skelter arrangements of things as earth's atmosphere. Galaxies are in every shape from spiral to ball, resting in every plane, and moving in every direction. Some move toward us, some away from us, some even collide, and they are of all different ages. Some are in the process of forming, with others falling apart with age. This indicates there may be far more behaviors in space created by its atmosphere than collected and cataloged here.

The Universe Behaves
Like A Unified Field

As noted in the introduction, building on Descartes idea that stars were merely "vortices" in the substance of space, and Faraday's idea that stars are merely "knots" in the fabric of space, Einstein view was that stars are merely areas of s high concentration of energy in the substance space, with solid matter forming in its center. And like the others he believed as technology advances we'll eventually discover that everything in the universe, including space, is interconnected and interchangeable. Evidence today tends to confirm that conclusion:

Interconnection: Physicists using modern technology have discovered when an atomic particle is split, with each half going off in opposite directions, the two parts remain physically connected though space. If something happens to one half of the particle flying off in one direction in space, the same thing happens to the other half flying off in the opposite direction. This could only happen if the two parts are physically connected.

And this interconnectivity is not a totally new idea. Tesla and Marconi used the idea of interconnection to invent radio in the early 20th century. They theorized that, if an oscillator is tuned to amplify electronic vibrations that are then broadcast into space, an oscillator tuned to that frequency at enormous distances will oscillate exactly like that in the transmitter. They were right, and invented radio.

Interchangeability: We've known for a few hundred years that the atmosphere of earth will morph into physical things and physical things will morph into the atmosphere. The oxygen in the atmosphere will combine with iron to produce

31

physical rust, and trees produce oxygen that morphs into earth's atmosphere. Advancing technology reveals the same happens regarding the substance of space:

For example, physicists often report that when they disturb space in the laboratory something they call "virtual matter" momentarily appears, and then quickly disappears. And light vibrations disturb space causing it to emit photons that then disappear back into space as the vibrations pass. In other words, photons are probably not emitted by the light source, but are emitted by space, and morph back into space when the light vibrations pass.

And astronomers report that, when positive and negative matter collide in space, they morph into the substance of space and disappear. And, as noted, when space is superheated by the sun, it apparently morphs into plasma, and when plasma cools, it morphs back into space. So we not only have evidence that space can morph into matter, we have evidence that matter can morph into space.

If these reports are accurate, it appears that space is made up of a unique substance that, if sufficiently disturbed, breaks up into positive and negative energy. Then, if the two get sufficiently separated, they morph into positive matter and negative or "dark" matter. However, if the two parts don't get sufficiently separated, they quickly rejoin and morph back into space.

Space Behaves
Like A Trampoline

Observers often describe space as behaving like a giant trampoline. They explain that a huge star will depress the trampoline to cause nearby objects to roll toward the depression to mimic the motion of gravity. This is an interesting analogy, but shouldn't we be trying to discover *what* the trampoline is made of, and exactly *how* the behaviors of gravity and inertia are created? Unfortunately, while we use the analogy in establishment forums, no such investigation occurs.

Bacon gives the reason for this strange lack of interest. He notes that our unaided human understanding pays no attention to behaviors that happen all the time without fail. It will give the behavior a name, then lose interest and pass on. So, since space always acts like a trampoline, and gravity and inertia always occur, cosmology gives the behaviors names and passes on:

"In my judgment philosophy has been hindered by nothing more than this – that things of familiar and frequent occurrence do not arrest and

detain the thoughts of men, but are received in passing without any inquiry."

However, if the trampoline occasionally failed to work, and gravity and inertia weren't so infallibly dependable, if things occasionally flew off the earth, or satellites or planets occasionally fell out of orbit, we'd quickly call on our scientists to discover *what* the trampoline is made of, and *how* gravity and inertia occur; so we can either learn to control them with an anti-gravity and anti-inertia mechanism, or, at least, predict their behavior.

In Summary: If all of these physical behaviors we've collected and cataloged occur in space very much as they do in earth's atmosphere, and we know the behaviors in earth's atmosphere are caused by earth's atmosphere, how can cosmology and our establishment continue to ignore the new evidence and continue to assume space is an empty vacuum? Bacon again suggests an answer:

"The human understanding when it has once adopted an opinion (either as being the received opinion or as being agreeable to itself) draws all things else to support and agree with it. And though there be a greater number and weight of instances to be found on the other side, yet these it neglects and despises, or else by some distinction

sets aside and rejects; in order that by this great and pernicious predetermination the authority of its former conclusions may remain inviolate.

"And therefore it was a good answer that was made by one who when they showed him hanging in a temple a picture of those who had paid their vows as having escaped shipwreck, and who would have him say whether he did not now acknowledge the power of the gods – "Aye" said he, "but where are they painted that were drowned after their vows?"

How different it would be if astrophysicists using the scientific method were reinstated in establishment forums. Trained to look for both positive and negative evidence, and to give them the same weight, it's my judgment they would quickly find enough new evidence today from advancing technology to determine *what* space is made of and *how* it produces its behaviors, discover more of the laws governing the behaviors of our universe, and we'd be living in an entirely new age.

But, while this is mere speculation, there is one thing of which I'm most certain. If Newton, Faraday, Poincare, Lorentz or Einstein were to return, every one of them would be absolutely appalled to discover our modern establishment has banned discussion of their Unified Field Theory they worked so hard to develop, and banned the ether dispute from discussion in its forums. They would all completely agree with Einstein:

"The free, unhampered exchange of ideas and scientific conclusions is as necessary for the sound development of science, as it is in all spheres of cultural life."

Of course, this open exchange of ideas by qualified people is the most basic premise of modern science. So the question arises, after all the progress brought about by astrophysicists working with the Unified Field Theory and participating in the ether dispute, precisely what could have happened a half century ago to cause our establishment forums to suddenly sweep it all under the table?

Those who lived through the "red scare" of the Cold War remember how it happened, but because it may be difficult for those who didn't live through the period to believe, let me take a few pages to expand on the well documented facts. Not only is it an interesting tale of medieval human behavior, it is also important to recount those facts in detail because they hold an important clue to the undoing of the damage.

Part II

The Medieval Reason
Establishment Forums Were
Closed To Astrophysics

From the dawn of the Age of Science in the 17th century, until the "red scare" in the mid-twentieth century, our establishment provided an open forum for both cosmology and astrophysics to discuss their theories. Cosmologists speculated upon *where* the universe came from, *where* it would end up, and *how many* universes exist; and astrophysics speculated upon *what* the universe was made of, *how* it created its behaviors, and *the laws* that governed its behaviors.

And, remembering the Galileo incident, both sides adhered to a tacit agreement to refrain from public criticism of the other's theories, with both tacitly ignoring the fact their views about the nature of space were in conflict. But just after the death of Einstein, astronomer Fred Hoyle assumed the mantle as *ad hoc* spokesman for astrophysics and unfortunately, lacking the tact of an Einstein, he seriously rocked the boat.

Hoyle Breaks Tacit Agreement: An ebullient Yorkshire man, host of a popular BBC radio program, and ably traveling about the world promoting seminars in astrophysics, Hoyle violated the tacit agreement between cosmology and astrophysics. Instead of working quietly to determine *what* the universe is made of, and *how* it creates its behaviors, and staying out of the turf of cosmology, Hoyle invaded the turf of cosmology, and began promoting a theory of *where* the universe came from.

Arguing there was no evidence the universe had a beginning or will have an end, and with only sketchy evidence of his own, he argued the universe is eternal, with stars and planets constantly being created and destroyed. He called the theory the Steady State Universe, but unfortunately often referred to the universe as an "evolutionary" process. The very word, of course, waved a red flag in the face of theologians and cosmologists.

Already organized to battle Darwin's evolutionary theory of *where* man came from, theology and cosmology suddenly found themselves confronted with a new evolutionary theory of *where* the universe came from. They began looking for a weapon to wage a counterattack, and quickly found the weapon in a theory of cosmologist and astronomer Edwin Hubble.

Hubble had noted that very distant galaxies exhibited a red shift that could indicate the galaxies were speeding rapidly away, or it could indicate that space affects the frequency of light over large distances. But ignoring the contrary evidence, Hubble argued the red shift indicated that biblical scholars were right. A single explosion had created every galaxy in the universe and the modern Big Bang theory was born.

When Hubble related his theory to Einstein, Einstein characteristically responded that it was an interesting idea, but quickly added, "It isn't physics!" To Einstein any theory of *where* the universe came from was the exclusive turf of cosmology, and irrelevant to the goal of astrophysics and its goal of determining *what* the universe is made of, and *how* it creates its behaviors.

However, as physicist Thomas Gold later remarked: "The biblically religious people wanted a moment of Creation, and obviously Hubble's Big Bang was their stuff." Georges Lemaitre, both an ordained priest and cosmologist, apparently got Pope Pius XII to publicly state Hubble's theory was "consonant with the Bible," and a formal clash between cosmology and astrophysics began.

In a well-orchestrated release, cosmologists notified English newspapers that "science" had just discovered evidence confirming the biblical story of The Creation. The statement wasn't at all true. Astrophysics wasn't connected with Hubble's Big Bang Theory, and, as Bacon had noted in the *Novum Organum*, modern cosmology had been promoting a theory based on Genesis long before modern technology and the age of science.

However, editors of English newspapers weren't interested in such details, and went bananas over the news. They saw a brand new controversy like Darwin's evolutionary theory dropped into their laps; only in this one they had a popular radio personality arguing "evolution," and Hubble and Lemaitre arguing "creationism."

Recognizing such controversy sells papers, they gleefully fanned the flames. The British Evening News blazed the headline: **"Science has proved the Bible was right."** The Evening Standard followed with the headline: **"'How it all began' fits in with the Bible,"** and local papers all over England quickly copied and jumped into the fray. In America, the press joined with equal gusto.

But the establishment, to its credit, kept relatively calm. As it had with the Darwin controversy, it dutifully provided a forum for both sides. One journal ran an article quizzing readers as to which theory of the universe they believed was correct, the "steady state" or the "creationist," and the response split down the middle. However, significantly, another question asked readers if they thought the first question was scientifically relevant, and got a 90% negative response.

The controversy probably would have eventually cooled down, as it had with the Darwinian matter, and the parties would have restored the tacit agreement and returned to tending their own turf and progress continued. But fiery Hoyle wouldn't back down. Like the newspaper editors, he was in showbiz, and apparently finding the controversy improved his radio rating, he pressed the fight

Unfortunately he made the mistake of Galileo. He argued with ridicule. His group often referred to cosmology as "cosmo-mythology," and made fun of cosmology's theory by being the first to derisively dub it "The Big Bang" theory. But then something happened overnight that would ultimately lead Hoyle to the same fate as Galileo, and unfortunately he would take the Unified Field Theory and the ether dispute with him.

Hoyle Ends With A Big Bang: One morning, in the midst of all his success, Hoyle awoke to the news that Klaus Fuchs, a casual associate, had been arrested for passing atomic secrets to the Russians. Shortly thereafter the Rosenbergs were arrested in America on a similar charge. The western political establishment reacted with panic. Fuchs was imprisoned in England, and the Rosenbergs were publicly executed in America.

Then a rumor began to circulate that Hoyle's "evolutionary" theory, which conflicted with Genesis, was a plant of "godless communism" designed to undermine the Christian west. We are left to guess who was behind the rumor, but, suddenly, everyone and anything connected with Hoyle, including the Unified Field Theory and the ether dispute, became tainted.

The western establishment went into a state of panic. With willing spies everywhere, the mere rumor of communist connection meant a summons before Congressional tribunals, and, like Galileo, subjection to public ridicule and a destruction of professional careers. Insane things happened. Physicist J. Robert Oppenheimer, for example, chief developer of the atom bomb, lost his clearance to atomic secrets.

Hoyle escaped direct accusation, but he suffered ridicule, lost his radio program, and establishment forums began ignoring him and anyone else pushing the theory of astrophysics. Thereafter, to be published, papers had to support the Big Band Theory of Genesis. Those supporting the Unified Field Theory or referring to space as the ether had little chance of publication. Cosmologists became gatekeepers of establishment forums, and astrophysicists, finding themselves excluded, assumed the role of mechanics designing space voyages.

And, today, 20 years after the end of the Cold War, the ban remains in effect, and it could go on forever until some emerging society – especially one not fascinated with Genesis but reading Newton and Einstein – creates a breakthrough by discovering *what* space is made of, and its interrelationship with energy and matter, and America could be left behind like Italy and Spain were in the 17th century.

The Silence of Astrophysicists: One might wonder, with astrophysicists working every day charting the physical topography and behavior of the great ocean of space, why some don't publicly speak out that space doesn't act like an empty vacuum? Why do they act like medieval chart makers who remained silent about the behavior of the universe until lawyer Copernicus spoke out? Francis Bacon attributed this silence to a working class mentality.

"The mechanic, not troubling himself with the investigation of truth, confines his attention to those things which bear upon his particular work, and will not either raise his mind or stretch out his hand for anything else."

But I think he let his class prejudice get in the way. Spanish and Italian chart makers and explorers certainly knew that biblical cosmologists had been mistaken about the shape of the earth and its position in the solar system, but with their establishments openly Defenders of the Faith, they knew, if they contradicted biblical cosmology, they'd wind up toast.

I'm sure it's the same with today's astrophysicists. If they have any ideas that conflict with the Big Bang Theory, they keep silent. They don't want to end up like Galileo or Hoyle. They need to keep their jobs and food on their tables. But, if we can't depend upon astrophysicists to speak up, how can we get astrophysics reinstated?

If Newton were alive today, he'd suggest we use the solution Francis Bacon proposed in his *Novum Organum.* Newton used it successfully to smuggle science into a 17th century English establishment totally dominated by scholastic philosophy, and it been used ever since whenever science has been threatened or pushed aside.

Part III

Bacon

"Let there be two streams and dispensations of knowledge, and in like manner two tribes or kindred of students in philosophy – let there in short be one method for the cultivation of existing knowledge, and another for the invention of knowledge."

Francis Bacon: *Novum Organum*

The Baconian Solution

Francis Bacon, Attorney General of England, Chief Judge of the English Supreme Court, an English Viscount, a philosopher, and, we need to add, a Machiavellian politician, recognized that inventions like gunpowder, printing, and the compass had changed the course of human history, and he set out to determine what led to such inventions and change.

Two Different Methods: Bacon discovered there are two different methods of inquiry. There is the *innate* method hardwired into the brain of every animal by evolution to provide survival. To survive an animal needs to know *where* his prey comes from, *where* it is going, and *how many* exist. And the quicker the animal can come to a conclusion, and not allow doubt to interfere, the more chance of a successful hunt and survival.

However, to move beyond mere survival and invent new things and progress, Bacon found one must take his time to gather all the facts, and then inquire into *what* things are made of, *how* their behaviors are created, and *the laws* that govern the behaviors, and, finally, test conclusions with experiment. With this unnatural method, man learned, for example, to domesticate animals to insure a steady supply of food and trigger the rise of civilization.

There's no doubt in my mind that Bacon came to his conclusions and wrote the *Novum Organum* from his experience as a sitting judge using the new "rules of evidence" then being developed and formalized in the English courtroom to improve the level of justice in England. His *Novum Organum* is full of terms and methods reflecting those rules.

The striking similarity between the method of inquiry used in a modern court system and in science may be the reason so many people trained in the law like Bacon, Copernicus, and Descartes have been so valuable to the early development of the physical sciences. And the reason lawyers like Jefferson, Adams and Madison were so valuable in the early development of political science. It's certainly the reason a lawyer is writing this book.

At any rate, it was the observations set forth in the *Novum Organum* that, in the hands of experimental genius like Isaac Newton, evolved into the modern scientific method. The difference between the courtroom method and that used in the science lab is important. In the courtroom a judge or jury makes the final conclusion, and in the lab conclusions are made by nature with experiment. Science is more successful because it isn't subject to the same degree of human fallibility as court trials.

Science Smuggled Into Society: From the beginning Bacon recognize it wouldn't be easy to get an establishment to allow the new method of inquiry into its forums. He knew from experience that most people will naturally continue using the innate survival method hardwired into the human understanding. They will find the new scientific method "lacking common sense," just as they found the use of the new rules of evidence in his courtroom.

And, since most people will continue to use the innate method, and the method is designed by nature to preserve the *status quo,* he recognized most people will find it difficult to even imagine progress and invention until after they appear:

> *"For when a man looks at the variety and beauty of the provisions which the mechanical arts have brought together for men's use, we will certainly be more inclined to admire the wealth of man than to feel his wants."*

And experience proved Bacon correct. The majority of people even today are suspicious of people who talk about change and progress. This is not from lack of intelligence, but because they are using a different method of thought than people talking about change and progress. Listen to Einstein lament this problem that, despite his fame, still dogged him four hundred years later:

> *"We must not conceal from ourselves that no improvement in the present depressing situation is possible without a severe struggle; for the handful of those who are really determined to do something is minute in comparison to the mass of lukewarm and misguided."*

The Dual Solution: Because of this propensity of people to remain conservative and be suspicious of change, Bacon thought it futile to try to replace existing scholastic philosophy with theories of the new science. Instead he suggested the new science be introduced into society as "handmaiden" to existing philosophies; suggesting, whenever a conflict arises, scientists treat established views with circumspect:

"If the matter be truly considered, natural philosophy is after the word of God at once the surest medicine against superstition, and the most approved nourishment for faith, and there she is rightly given to religion as her most faithful handmaid, since the one displays the will of God, the other his power."

In short, Bacon suggested the creation of a *dual* system of philosophy. Both under the protection of the establishment, scholastic philosophy would be left in place, and the new science would work quietly in the background. He felt confident, if both were protected, given the facts and time, society would eventually absorb and accommodate new inventions at its own comfortable pace and even hold science in awe:

"Such is the infelicity and unhappy disposition of the human mind in this course of

invention, that it first distrusts and then despises itself: first cannot imagine that such thing can be found out; and, when it is found out, cannot understand how the world should have missed it so long."

Targeting Movers and Shakers: Lastly, and most important, Bacon knew who to target to help implement this dual system. He realized it is of no use to try to convince scholastics philosophers of the value of science or urge them to accept it as a valuable partner. "It cannot be that we should think alike, when one drinks water and the other drinks wine."

Instead, his experience in English politics had taught him that everything in society eventually boils down to economics and power, so he dedicated his *Novum Organum* to the primary movers and shakers of his 17th century society – The King and the English aristocracy.

And he never missed a chance to promote the new science to the financiers, shipping magnets, military leaders, and church fathers among his peers to convince them it would father new inventions that would increase the wealth and power of England, and, most important, of themselves. History proves Bacon targeted exactly the right people.

Sir Robert Boyle

The British Gentry Adopt Bacon's Dual Solution

In 1660, shortly after Bacon's death, a group of English aristocrats like Sir Robert Boyle, joined by merchants, industrialists, financiers, military men, and some theologians seeking dominion over more souls, became convinced that Bacon's new scientific method could make England and themselves wealthy and powerful. They petitioned King Charles II to grant them a charter to organize a formal scientific society.

The King, anxious to replace Spain's current hegemony over the seas, and make his realm richer and more powerful, chartered, *The Royal Society of London*; the first formal organization to be dedicated to the scientific method, declaring himself a charter member. The members then dedicated the new society to the memory of Francis

Bacon, and a new Age of Organized Science was formally launched.

In 1672 Isaac Newton became a Fellow of the society and in 1703 its Director, and remained its virtual dictator for another thirty productive years. Educated in a seminary and familiar with theology and cosmology, Newton set up what was to be a long term *tacit* agreement between science and religion, wherein the new science would be sold just as Bacon suggested, as handmaiden to religion.

It proved a highly successful arrangement. Unlike Italy where Galileo had used the work of Copernicus to initiate a Renaissance of Science, but had insulted the Church of Rome with a book portraying the protagonist of biblical cosmology as a simpleton, the Royal Society under the leadership of Newton treated biblical cosmology with deliberate circumspect.

And the king, his aristocrats, and the financiers, military and theologians reaped the harvest they expected. With the increase of wealth and power provided England by the new organized science, English kings were to reign over the largest empire in history, while the aristocrats, financiers and military grew rich and immensely powerful, including The Church of England that ultimately was to hold dominion over as many souls as The Church of Rome.

Darwin Breaks Tacit Agreement: However things didn't go well after Newton. Without his genius in control, the Society allowed member Charles Darwin to upset the apple cart. On a voyage financed by the society, Darwin discovered that animals developed new physical qualities through "survival of the fittest." The observation and his *Origin of the Species* led to the modern science of genetics.

But Darwin didn't stop there and carelessly made the mistake of Galileo and, later. Fred Hoyle. Darwin invaded the turf of theology by writing another book about *where* man came from, arguing that man was not made in the image of God, as the biblical theologians reported, but was descended from monkeys. As you can imagine, the book set off a battle between theology and science that still rages.

However, western establishments played it cool and continued to report both the argument of theology and of science on the subject in its forums. But in the area of politics, the English establishment stumbled badly. It failed to provide an open forum for the new political science developed by Englishman John Locke, and ultimately paid the price.

The old system of politics based upon the scholastic method of inquiry had granted power to people based upon *where* they came from, and like all systems using this scholastic method, its political system wasn't leading England to political progress. John Locke's new system proposed to allow the public to elect politicians based, not on *where* they came from, but on how they proposed to behave politically and economically.

If we examine political science, it's as revolutionary as modern physics, but the English establishment, instead of providing this new branch of science access to it forums, banned discussion of Locke's theories from its forums. The result was England ultimately forfeited political leadership and the power and wealth it provided to its emerging American colonies where the ban was not in force.

Jefferson

"Three of the most important people who ever lived are Bacon, Newton, and Locke."
Thomas Jefferson

The American Gentry Adopt Bacon's Dual Solution

Shortly after Locke's death, the economic and political leaders of America – its lawyers, planters, financiers, businessmen, shipping magnets, military people, and theologians – convinced that Locke's new scientific political system could help improve the wealth and power of the colonies and themselves – set out to experiment with it in America.

They ultimately adopted the new American Constitution that included a dual system commonly known as "separation of church and state" that guaranteed an open forum for both theologians and political scientist to express their divergent views on politics and religion. Theologians were free to push for a return to theocracy, and political scientists were free to push for an expansion of secular democracy.

In practice America has generally split into about half the population continuing to use the old survival method, seeking to protect the *status quo* in politics and economics and remaining suspicious of progress; and the other half using the new method of thinking seeking to bring change and progress. On the surface to many this seems a shame because it foments arguments and prevents decisiveness.

However the duality has worked out rather well, with constructive conservatives providing stability to the nation's politics, and constructive liberals pushing for political experiment providing progress. In fact the experiment with this duality has been such a rousing success that every modern state has adopted the duality in some form, and societies that haven't lag behind.

But the duality in America is tenuous. In times of war and fear, the American establishment will often ignore its Constitution, and close its forums to open discussion and politically and economically retrogress. However, in most cases, as soon as immediate danger passes, it will reluctantly restore the duality, and prompt a renaissance of political success and progress.

But regarding astrophysics, this hasn't happened. The "red scare" of the Cold War ended 20 years ago, but our establishment forums, it could be argued in violation of the American Constitution, still categorically reject articles supporting the Unified Field Theory and ether dispute for no other actual reason than they conflict with Genesis.

So, while other sciences, enjoying duality, have all progressed so much in the last half-century, they don't seem the same sciences; there has been no such progress in astrophysics. We know little more today about *what* the universe is made of or *how* it creates its behaviors than we did a half-century ago – solely because we're not looking for progress.

And, if we find it hard to believe that four centuries into the age of science our establishment could be guilty of such medieval behavior, we need only remind ourselves that Francis Bacon warned of such periodic regression. He observed from experience that the human understanding will often tire of the new method of inquiry, and long to return to its old scholastic ideas:

"The idols which are now in possession of the human understanding, and have taken deep root therein, not only so beset men's minds that truth can hardly find entrance, but even after entrance obtained, they will again in the very instauration of the sciences meet and trouble us, unless men being forewarned of the danger fortify themselves as far as may be against their assaults . . .

"All (idols) must be put away with a fixed and solemn determination . . . for the entrance into the kingdom of man, founded on the sciences, being not much other than the entrance into the kingdom of heaven, where into none may enter except as a little child."

If we don't heed Bacon's warning and put away our dependence upon biblical cosmology to understand our universe, we are placing our leadership of this Age of Science at risk of being lost to new emerging societies that are not fixated with our religious idols and received opinions. We may awake one day to find our leadership of the Space Age snatched from us along with the wealth and power it provides.

If Bacon, Newton or Einstein were here today, they'd suggest us citizens, who now have the vote, take part in restoring the duality. They would list three simple steps their experience shows would help reopen establishment forums to astrophysics, and create a *renaissance* in the science of understanding our universe.

Part IV

Three Proven Steps to Reopen Establishment Forums

1. **Formally separate cosmology and astrophysics:** Our establishment must recognize that cosmology is not a science, but is a scholastic philosophy; and formally place it in the department of philosophy along with theology and other scholastic disciplines. Cosmology would be charged with using our natural method of inquiry as outlined in Aristotle's *Organum* to speculate upon the unknown – to provide satisfying theories of *where* our universe came from, *where* it is likely to end up, and *how many* universes exist.

At the same time the establishment must recognize astrophysics is a behavioral science, and place it in the department of science along with medicine, particle physics and political science. This science would be charged with using the scientific method to satisfy our need to know *what* the universe is made of, *how* it works, and the *laws* that govern its behavior so we can resume progress in understanding the behavior of our universe.

2. Restore the tacit agreement between cosmology and astrophysics: Astrophysicists must learn the question of *where* the universe came from is the exclusive turf of theology and cosmology, and is irrelevant to science's goal. Astrophysicists need to avoid Hoyle's mistake of arguing with theology and cosmology about *where* the universe came from, and stick to the job of quietly determining *what* the universe is made of, *how* it works, and *the laws* that govern it.

On the other hand, cosmology must remember biblical cosmologists were wrong about physical aspects of our universe. The earth is not flat nor the center of the universe; and, likewise, space may still prove not to be the empty vacuum it appears. So cosmologists need to play down their view that space is empty vacuum. In fact their Big Bang theory would be more easily acceptable if they argued the galaxies were created out of the substance of space.

Both astrophysicists and cosmologists need to remember the sage view of Einstein: *"Science without religion is lame. Religion without science is blind."* The point is the Big Bang theory can't replace the Unified Field Theory, and the Unified Field theory can't replace the Big Bang theory. They are apples and oranges. The Big Bang theory is about *where* the universe came from, and the

Unified Field Theory is about *what* the universe is made of.

As for *the ether dispute*, it is a question of fact that will eventually be scientifically resolved by advancing technology.

3. Target the movers and the shakers of society: Finally, they would tell us keep in mind Bacon's observation that everything boils down to economics and power. We don't need to convince cosmologists or astrophysicists of the need to reinstate duality. That attempt would only foment an argument and end in deadlock. Instead, we need to convince the movers and shakers of our society – our economic, political, military leaders, and religious leaders – of the economic and military value of a dual approach.

History shows these are the people who quietly control our society. With a casual word from them in their many contacts, we'd soon discover independent chairs of cosmology and astrophysics popping up quietly in our universities, and establishment forums once again quietly accepting papers from both qualified cosmologists and astrophysicists, assuring us of our place in The Space Age, and the wealth and power (and pride) it brings us all.

Bibliography:

Novum Organum: *On the Interpretation of Nature*, 1620, Francis Bacon, Spedding Edition, Random House

Einstein: *His life and universe:* 2007 Walter Isaacson, Simon and Schuster

Fred Hoyle's Universe, 2005, Jane Gregory, Oxford Press

The Big Bang Never Happened, 1990 Eric Lerner, Random House

www.ingramcontent.com/pod-product-compliance
Lightning Source LLC
Chambersburg PA
CBHW051242170526
45165CB00004B/1544